Why Matl

An exploration of r

By Daniel Cove

To 11T2, the best tutor group a teacher could hope for

Introduction - What is Mathematics?

I have an absolute passion for maths. I studied it for five years and have now been teaching it for a further five, and in that decade my love and interest for the purest of subjects has only ever grown. I know that many people do not enjoy maths (and have encountered more than my fair share), and many people believe they simply "cannot do maths". It strikes me as an odd assumption, as I have never heard anyone assert that they "cannot do English", but that's a discussion for another time. To discuss why maths is interesting, we must first establish what maths actually is.

When most people think of maths, they think of adding numbers. They think of multiplication, possibly long division, and even the dreaded fraction. I've seen that word strike fear into more hearts than probably all others combined. But to me, these things are not mathematics. They are simply tools with which maths is done. This is akin to asking what poetry is, and having the alphabet listed at you. The alphabet is not poetry itself, merely a tool which allows poetry to be created.

There are various definitions of maths, many too tedious to list, but they all have a common theme – the idea of patterns, rules and proof.

Firstly, finding patterns in numbers and objects. Finding rules which take the knowledge from one scenario and places it into another scenario. For example, mathematicians studying how spots grow on leopards were able to apply the same rules to preventative forest fire measures, as fire in a forest and chemicals in skin disperse and react in mathematically similar ways. To me, this is not only interesting but clearly useful.

Secondly, the idea of proving something. Mathematicians are never content to say 'close enough'. Maths is advanced only by those who show that a statement is true beyond all question, by examining every possible nook and cranny of its being until its full meaning is understood, examined, and proved true or false. One of the most famous examples of this is Fermat's Last Theorem. Fermat was one of the greatest mathematicians ever to have lived, and back in 1637, Fermat made to be what seemed to be a fairly modest claim: that $a^n + b^n = c^n$ has no whole number answers when n is bigger than 2. He made a note in a book that said he 'has a proof which is too large to fit in this margin', and promptly died without ever writing it down. (He actually lived for another 30 years, never writing it down, and the note was only discovered after his death). Cue mathematicians for the next 300 years trying to establish whether it was actually true. It seemed true. Nobody could find an example where it wasn't true. But to actually *prove* that it was always true for every single possible combination of numbers turned out to be a colossal undertaking. Like a huge jigsaw puzzle, mathematicians worked tirelessly to slowly put it together. Each new insight gave a new piece, a new idea to work from, until finally it was proved in 1995, a full 358 years after the original claim was made.

So what drove dozens of mathematicians to dedicate years of their life to this problem? In my mind, it's the same reason why collectors strive to own everything from a particular set, or why some people meticulously search their family trees for information. They want to leave no stone unturned, they want to have a sense of completeness. Proving a statement in maths is the ultimate way to do this, and may well be what drives mathematicians of the past and present.

So that's what maths is – a series of connections, rules, patterns and proofs that somehow interlink like a huge web to give us a rich field of study. A field of study which allows us, from fundamental rules, to create new structures, explore new methods, solve new problems and even invent new numbers. I have taught many students who simply cannot conceive of this – several imagine that professors of maths spend their time adding very big numbers together, or even that they spend their time trying to find bigger numbers than each other. For me, a mathematician is anyone who can find rules, apply logic and view a situation or problem in more than one way.

To understand why some people do not like maths, I use an analogy from Professor Edward Frenkel - imagine you had to take an Art class at school, but for years were only ever taught how to paint a fence. You were never shown the works of the great masters, never allowed to see the paintings they made. You just painted that fence. Would that make you an art lover? Of course not. Years later, you might say to your friends that you hate art; but what you'd really be saying is that you hated painting that fence.

A very similar situation arises with maths. In the real world, not everybody visits art museums, but everyone knows that they are there, and if they have an interest in Art they can visit them. With maths, most people are not even aware of the masterpieces. Maths is not part of our cultural discussions in the same way that science and art are, and so in general the lack of awareness of its rich, powerful world leads people to believe that it is boring, useless, or both.

I hope that's given you some perspective on maths as a whole – in the following sections I will be exploring various mathematical ideas, methods, tidbits and trivia. Some will entertain, some will intrigue, some may even frustrate, but ultimately all will show you how and why I find maths so very, very interesting.

Stairs, rabbits and galaxies

Let's imagine that in your house, you have a flight of 12 steps. Your legs are long enough that you can go up either 1 step at a time or 2 steps at a time. For example, you might choose to go up 1,1,1,2,2,1,2,1,1 as shown below.

This poses the question: in how many different ways can you go up this staircase?

This seems a fairly innocuous question, and the first instinct of most people is to start listing all the possible ways: 1,1,1,1,1,1,1,1,1,1,1,1. Then 1,1,1,1,1,1,1,1,1,1,2. Then 1,1,1,1,1,1,1,1,1,2,1. At this point, many realise that this will take a *long* time to list all of the possible ways. Even noting that there are 11 different ways to fit a single '2' in the list, progressing to having all combinations of two '2's in the list is a huge task. How could you make sure you listed all of them? After listing 1,1,1,1,1,1,1,1,2,2, should you list 1,1,1,1,1,1,1,2,2,1 or 1,1,1,1,1,1,1,2,1,2 ? There doesn't appear to be a quick of way working it out, and we're still nowhere near answering the actual question.

So, we do what any mathematician would do: break it down into the simplest possible problem, begin to build it up, and look for patterns and rules.

The simplest possible problem is to only have one step in your staircase. In that case, there is only one way to go up the steps − 1.

What if there are two steps? Well, then there are two ways – 1,1 and 2. Three steps isn't too tough to list either, giving three ways. 1,1,1 or 1,2 or 2,1.

You might be tempted to claim that four steps must therefore have four ways, but it actually yields five: 1,1,1,1 or 1,1,2 or 1,2,1 or 2,1,1 or 2,2. Five steps has eight different ways of 1,1,1,1,1 or 1,1,1,2 or 1,1,2,1 or 1,2,1,1 or 2,1,1,1 or 1,2,2 or 2,1,2 or 2,2,1 (these are starting to get annoying to list now) and from here we can start compiling our results to look for patterns and rules.

Number of steps	1	2	3	4	5
Number of ways	1	2	3	5	8

Is there a rule that would enable us to work out how many ways there are to walk up six steps? Look at the 'number of ways' row, and we can see that each result is simply the previous 2 results added together. 1+2 = 3, then 2+3 = 5, and 3+5 = 8. So to work out the number of ways for six steps, we simply add 5+8 to give us 13 ways. (At this point, we haven't *proved* our rule works all the time, and it would take much more maths and many more pages, so for now we'll just take it as fact).

Continuing the pattern gives us these results:

Steps	1	2	3	4	5	6	7	8	9	10	11	12
Ways	1	2	3	5	8	13	21	34	55	89	144	233

This gives us the final answer to our problem, that there are 233 different ways in which to climb our stairs (thank goodness we didn't try to list all of them).

But that's not where this ends. Some readers may recognise the sequence 1,2,3,5,8,13,... in the 'number of ways' row. They are very famous numbers in the mathematical world and known as the Fibonacci Numbers, discovered in 1202 by the mathematician Fibonacci whilst he was living in Italy. But he didn't discover them while talking about going up a flight of stairs – he was concerned with a completely different problem about rabbit populations.

Imagine you have a pair of adult rabbits. Each month, they produce a new pair of baby rabbits. It takes one month for baby rabbits to mature, at which point they will also start producing a pair of baby rabbits each month. After one year, how many pairs of rabbits will you have?

Using A to mean Adult and B to mean Baby, each month A becomes A,B (as they have reproduced) and any existing B become A (as they mature).

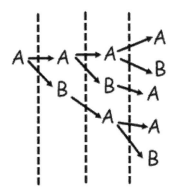

Putting this into a flow diagram as above shows us that the number of pairs each month is 1 (A), 2 (AB), 3 (ABA), 5 (ABAAB) and so forth, giving us the Fibonacci sequence of 1,2,3,5,8,13,… meaning that after a year (i.e. 12 months) there will be 233 pairs of rabbits.

So two seemingly different problems have identical solutions – but it doesn't even stop there.

Fibonacci went on to use his sequence directly to draw out the so-called Golden Spiral, created by taking each number in the Fibonacci sequence and drawing it as a square:

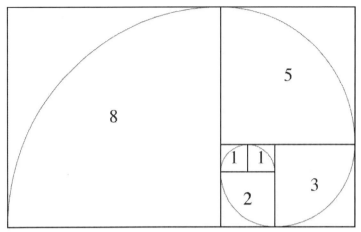

This spiral turns out to be the shape of many sea shells, human ears, hurricanes, and even our galaxy.

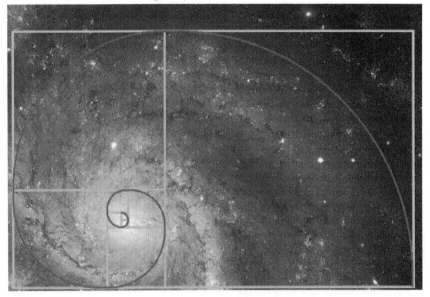

We started out with a problem that involved a person walking up a flight of stairs, and it turns out that exactly the same maths we used to solve it tells us the shape of our galaxy. What other subject could make such a claim? What other area could make such a leap?

And this is just one of the many connections there are in maths - we have used just one of the thousands of rules and patterns that can be found.

Never before in history – in your house

In your house, probably right now, you own something that no other human has ever possessed in the entire history of mankind. I actually have two such objects sitting on the shelf quite near me (each different from the other, of course). What is that object? A standard deck of 52 playing cards. Assuming it's been well shuffled, then the exact order that your deck is in is unique. No other deck of cards in the entire history of mankind has ever been in that particular order. (Again, we have to assume that the decks of cards have been well shuffled. You could of course get two decks of cards and laboriously place them in the same order as each other – or, more easily, buy two new decks of cards).

How can I be so sure? Well, first we have to get a sense of how many ways there are to rearrange a deck of 52 cards. There are 52 choices for the top card, then 51 choices for the next card, then 50 for the next card, and so on. This gives us 52 x 51 x 50 x 49 x …. x 3 x 2 x 1, which in maths can be written as 52! to save space. How large is this number? It's : 80,658,175,170,943,878,571,660,636,856,403,766,975,289,505,440,883,2 77,824,000,000,000,000 which is extremely large.

Finding an analogy to describe the size of this number is difficult. You might start by trying to compare it to, say, the number of grains of sand on a beach. Or even the number of grains of sand on planet earth. Well, the number of grains of sand on planet earth is estimated to be 7,000,000,000,000,000,000 which to be honest isn't even close. Our number, 52!, is trillions and trillions of times larger than that.

There aren't many things that I can easily imagine that are smaller than a grain of sand, so let's perhaps think how long 52! is in terms of time. Let's think about 52! seconds and how long that would be. 100 years? Probably longer. 1000 years? 1,000,000 years? Well, let's put it in some context. Let's say you measure time in an odd way. Once per second, you take one drop of water, anywhere on earth (any of the rivers, oceans, anywhere) and fling it into space. Every second, one drop of water disappears from somewhere on earth. Now to empty the whole earth of water is going to take a LONG time. Think of the vastness of the Pacific Ocean alone. The amount of water you need to dispose of is colossal. Would you be able to empty the entire earth of water, one drop at a time, in 52! seconds? As it turns out, yes you can. And not just once. You've got time to start refilling the earth as well, one drop at a time. One drop per second, you can start refilling the entirety of earth's water. In your 52! seconds, you can not only drain and refill the earth of water, you can do the entire process 3,000,000,000,000,000,000,000,000,000,000,000,000,000 times.

In some senses, 52! is a number so large that it's difficult to even understand with analogies. But assuming that our brains can cope with how large that number is, we return to decks of cards (remembering that 52! is how many different ways there are to rearrange a deck of cards).

To show that no two well-shuffled decks have ever been in the same order, we do what mathematicians call finding an *Upper Bound*. Let's make some outrageous assumptions about the history of humanity, find the probability that any 2 decks of cards ever matched under those assumptions, and then we'll know that the actual probability is far, far less than that.

Let's assume that the population of earth has been 7 billion people since the big bang (which it definitely hasn't). Let's also assume that since the big bang, each and every one of the 7 billion humans has been shuffling a deck of cards every 5 seconds, without ever pausing for sleep (which they definitely haven't). These are obviously ludicrous assumptions to make.

Now the big bang happened approximately 14 billion years ago, and each human is shuffling 6,307,200 new decks each year – multiplying 14 billion x 7 billion x 6,307,200 will give us the total number of new decks created over the course of this humanity, which is approximately 618,000,000,000,000,000,000,000,000. While this is a very large number, it's nowhere near 52! (take a look back to see the difference in size). It's still millions and millions of times smaller. Even under those crazy assumptions, the probability of any 2 decks of cards ever matching is a number so small that every calculator or online number cruncher I can find tells me it's 0%. These things tend to have a capacity of 100 digits, so that means, at most, the chance of any two decks of cards matching is 0.000...0001% where there are at least 99 zeroes before the 1. That's less likely than me, you, your 10 closest friends, David Cameron and all five members of the Spice Girls all being struck by lightning on the same day. And that's under all the ridiculous assumptions we made. The actual probability is FAR less than that.

And that's how I'm certain that no deck of cards, in the entire history of humanity, has ever been in the same order as the deck of cards sitting in your house.

(Now strictly speaking, it's not actually impossible. There is an *unbelievably* small chance that two decks of cards have, at some point in history, been in the same order. However, that chance is so incredibly unbelievably small (see above), that we refer to it as statistically negligible; for the purposes of working things out, we just ignore it. It's so small as to be completely irrelevant.)

Some numbers don't exist

Think of a random number. Chances are, you thought of a positive whole number, like 13 or 8. But, as you well know, not all numbers are positive whole numbers (known to mathematicians as the Natural numbers). We can also have 0, and the negative whole numbers like -5 and -27, and between these two we have the all the whole numbers, known as the Integers.

Very rarely when asked to think of a whole number, someone will think of a fraction or decimal – something like $^2/_5$ or 1.8, which is yet again another way to extend the list of numbers which we already have into what are called the Rational numbers. Some of them are easier to express as decimals, some are easier to express as fractions, but it's important that it is possible to express them all as fractions. An easy example is that 0.5 = ½ . A trickier example is that 1.25 can be written as $^5/_4$. (If you don't believe me, enter 5 ÷4 on a calculator) An even stranger example is something like 0.3333333.... = $^1/_3$ or that 0.62626262.... is $^{62}/_{99}$.

This begs the question – are there any numbers in existence which do not fit into any of the categories above? It turns out that there are numbers which do not fit into those categories, and they are called the Irrational numbers. If you remember how to find the area of a circle, you will remember the formula $A = \pi r^2$, and may even remember that π is about 3.14 or 3.142, depending who you asked. Note my careful use of the word 'about' – because the exact value of π is almost impossible to write down. If you write it as a decimal, its digits never stop and never repeat into a pattern. Ever. It's an infinitely long string of randomly ordered digits. 3.14159265358..... Knowing one digit doesn't help you work out the next digit either. Because of these properties, it's completely impossible to write it as a fraction, making it an Irrational number. It turns out that there are many other (in fact, infinitely many) irrational numbers with this curious property that they cannot be written as fractions, and their decimals are doomed to continue forever without ever repeating into a pattern.

So now we surely must have found all the numbers? We have all the whole numbers, negative numbers, numbers that can be written as fractions, and numbers that cannot be written as fractions. So do we have all the numbers? Almost. We have found all the Real numbers. Why are they called Real numbers? Simply put, because some of the numbers we use in maths don't even exist. Let me explain.

What is the square root of 36? Most readers will recall that we are looking for a number that when we square it (i.e. multiply it by itself), we get the answer 36. It turns out that 6 x 6 = 36, so the square root of 36 is 6. At this point, it's worth noting that -6 x -6 = 36 too, (as two negatives multiply to make a positive) so actually -6 is the square root of 36 too.

We can go along merrily finding the square roots of numbers (it turns out some of them, like the square root of 20, are Irrational) until we stumble across this question:

What is the square root of -1? It's not 1, because 1 x 1 = 1. It's not -1 either, because -1 x -1 = 1 as well. Look as hard as you want, you'll never find a number that multiplies by itself to give you negative one (or, indeed, any negative answer). So what did mathematicians do? They defined the Imaginary numbers. The first imaginary number is called i and it has the property that $i \times i = -1$. Now you can't have i objects in your hand, but I can show you what it looks like.

Most people are familiar with the good old number line (only whole numbers shown):

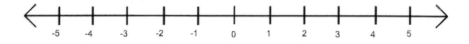

But what most people don't know is that there are actually two number lines – one for the Real numbers, and one for the Imaginary numbers.

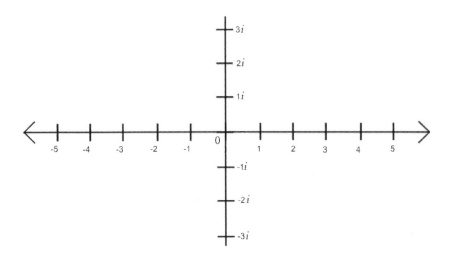

Going horizontally are the Real numbers as we know them, and then going vertically are the Imaginary numbers. The number lines only cross at the point 0 – because if you've got nothing, then you've got nothing, imaginary or not.

You'd be well within your rights to question whether mathematicians are allowed to do this, and what possible purpose they could serve. Even after their inception in the 1600s, some mathematicians held doubts about them until the early 1800s. In my mind, creating (or was it discovering?) the Imaginary numbers is just as valid as creating the negative numbers – you can't have -5 sheep in a field, you can't hold -5 things in your hand, but the number is right there on the number line. So long as they are defined using precise rules, then they are just as valid as any other number. Nowadays, negative numbers are commonplace in everyday life (for example many bank accounts have a negative balance) but when they were first introduced caused controversy amongst mathematicians – perhaps in 300 years, Imaginary numbers will be commonplace.

As for their uses, they are widely used in not only maths, but also biology, physics, electrical engineering and economics. For example, in electronics the state of a circuit can be described using Imaginary numbers, as can protective crash zones in cars and certain kinds of population growth. The actual maths involved is too lengthy to describe here, but suffice to say that they are widely applied and critical to the workings of much of what we know.

0.99999....

Continuing our theme of numbers that cause controversy, I have seen more debates and discussions about 0.9999... recurring than possibly any other mathematical topic – to the point where it even has its own Wikipedia page. The discussions (arguments) revolve around whether 0.999... is equal to 1. There are two distinct camps; those who believe that 0.999... and 1 are different numbers, and those who are right.

But before I jump into the discussion itself, it's worth pondering why this question has provoked so much discussion. Why do so many people have such strong opinions, and why do legions of people continue to debate back and forth? In my opinion, it's because most debate boils down to, and can centre around, what a number actually is. At its most fundamental level, what actually is the number 1? It leads to discussions about whether a number can be infinitely small, and what that even means (or could mean). It leads to discussions about what infinity is, and whether the string of 9s goes on forever, whether that's allowed, and what that means.

These discussions break down the very concept of numbers, arguably what is at the very heart of all mathematics. Without numbers, how can maths exist? Without an understanding of numbers, how can we understand maths? Without numbers and an understanding of them, what would maths even be? Would it be anything?

Mathematicians in the past have struggled with these very questions. Imagine measuring a number line, between the numbers 3 and 4. It's a total distance of 1. What if you removed the section between 3.5 and 3.6? Well then your total measured distance is only 0.9 (as 0.1 is missing). But consider this: what if you only remove the single point of 3.6 – what is your total measured distance then? Well surely a single point on a line has no distance, that's the very definition of a point. If it has any length, it would be a line. To claim that a single point had any length would be to claim that every single number had length; if every single number has a length, then the length of the whole number line is bigger than the length of the number line. That's clearly ridiculous, so each point must have a distance of 0.

In that case, if we only remove the point 3.6 from our section, then our section still has a total measured distance of 1 (as we have removed something that has a measured distance of 0). But if we're free to remove 3.6, we can just as equally remove 3.1 and 3.2 and 3.3 and 3.871 and 3.603 and 3.862983 and, really, any number we like. So we remove all of them, one at a time – yet each time we remove one, we're not altering the total measured distance as we're removing something which measures 0. Once we have removed all the numbers, we nothing left yet it still has a measured distance of 1. How can that possibly work?

Answering that question almost broke the whole of maths. It wasn't until Henri Lebesgue published a dissertation in 1902 (very recent in mathematical terms) which fundamentally redefined the precise meaning of the word measure, and did all sorts of things including creating sets of numbers which could not be measured, and different kinds of infinity. Using his work, maths was once again able to confidently use numbers, safe in the knowledge that we know how they work, and fundamentally, what they are. I'll explain more later in the book.

So, returning to 0.999... recurring, we must first agree on just one point; that in between any two numbers, there is another number. There are no two 'adjacent' numbers; if they were truly adjacent, then the gap between them would be 0, and if the gap between numbers is 0, then they are the same number. If the gap between them isn't 0, then I can find a number in between them.

Thus, when asking whether 0.999.... and 1 are the same number, I challenge you to find a number in between them. Remember that those 9s go on *forever*. They never, ever stop. Some people are tempted to say that there is a gap between them of 0.000...0001 with infinitely many 0s. This is wrong, because as soon as you write the 1, the 0s have ended and you haven't written infinitely many 0s. But even if this fictional number did exist, it would surely be the smallest number. There can't be a number smaller than 0.000...0001 when you write infinitely many 0s. But what happens when you want to halve that number? It's already the smallest number. You end up with a contradiction, and your only choice is to admit that 0.000...0001 doesn't really exist.

If that doesn't quite convince you, then think back to the difference between Rational numbers (which can be written as a fraction) and Irrational numbers (whose decimals never stop and never repeat into a pattern). 0.999... has to be either Rational or Irrational, so which is it? It can't be Irrational, because its decimal follows a clear repeating pattern, so it must be Rational. If it's Rational, you must be able to write it as a fraction – the only sensible answer is to write it as $^1/_1$ which is just 1.

Neither of these is a mathematical proof, just some convincing statements. A quick internet search will reveal many equally convincing, if flawed, arguments claiming that they are different numbers. There is a full, undeniable mathematical proof that they are the same number, which involves writing 0.99999.... as 0.9 + 0.09 + 0.009 + ... and doing some rather nifty work with that, but unfortunately it's too complex to give here.

At the very least, it gives me hope that a mathematical question has promoted so much discussion and, in one way, become part of our cultural discourse. Perhaps in the future more mathematical concepts and discussions will cross that tricky threshold and make their way into our daily discussions.

The infamous Monty Hall problem

Even if you don't know it by name, most people have heard, or heard of, the Monty Hall problem. I'll give the exact original wording, as first published in a letter written to *American Statistician* in 1975:

"Suppose you're on a game show, and you're given the choice of three doors: Behind one door is a car; behind the others, goats. You pick a door, say No. 1, and the host, who knows what's behind the doors, opens another door, say No. 3, which has a goat. He then says to you, "Do you want to pick door No. 2?" Is it to your advantage to switch your choice?"

This is a great question, and one which has generated much discussion. Before we hit the problem itself, it's worth noting that Monty Hall himself (a gameshow host) never, to his recollection, actually did this for any contestants. The problem was simply posed as a hypothetical question based on the TV gameshow called *Let's Make a Deal*, which was hosted by the eponymous Monty Hall. The choice of a goat as the non-prize appears to be random, but the wording has stuck, and even modern day wordings of the puzzle contain cars and goats (presumably on the assumption that very few people want to win a goat).

You may know the answer to the problem, or you may think you know the answer to the problem, or you may not know the answer at all. Regardless, I would encourage you to think about it now. It caused a huge stir when it was first published – as the magazine published the (correct) answer that you are more likely to win the car if you switch doors. It's a counter-intuitive answer, as most people (myself included when I first heard the problem) think that switching has no effect; that each door has a ½ chance of having the car. So counter-intuitive, in fact, that ten thousand readers wrote in, exclaiming that the magazine was wrong –one thousand of them with a PhD.

It's understandable too; the flawed logic that swapping makes no difference is easy to follow. Once the host opens a door, there are only two doors left, and so it appears that each door has a ½ chance of being the correct door.

So why is swapping the correct answer if it seems so obvious that swapping is inconsequential? Well, the first alarm bells should go off in your head if you imagine that instead of 3 doors, there are 1000 doors. Imagine a huge row of 1000 doors, with 999 containing goats, and only one with that car you so desperately want. You have to choose a door, so plump for door number 472 (come on lucky 472!). At this point, it's fairly clear that it's very unlikely you chose the right door. Monty Hall then opens door 1, showing a goat. He opens door 2, showing a goat. He continues opening every door (3,4,5,6,....) revealing a goat behind each and every one of them. He continues all the way opening doors until door 1000, but on his way, for reasons best known to himself, he doesn't open door 803. He leaves 803 closed.

He then offers you the choice of sticking with 472, or swapping to the mysteriously unopened 803. I think in this case, most people can see that swapping is a good idea. It's very unlikely that you managed to pick the car when you originally chose your door, which means it's very likely that the car was in 802 all along, which is why Monty left it closed. (Once in a thousand tries, you'll manage to accidentally pick the car first time and swap to a door with a goat behind it). So in this situation, swapping is definitely best – you only have a $1/1000$ chance of winning the car if you stick, as there was only a $1/1000$ chance that you chose the car in the first place.

In my head, this solution makes sense – all we have to do now is use this exact logic, but reduce the number of doors to three. You can work through the exact same procedure, but with only 3 doors, and the answer is the same (but just feels less intuitive). You only have $1/3$ chance of winning the car if you stick, because there is only a $1/3$ chance that you chose the car in the first place.

There's another way to think about this: You plump for lucky door number 1. At this point, it's unlikely that you have chosen the door with the car.

You only have a $1/3$ chance of being right. It's much more likely ($2/3$ chance) that the car is somewhere else, i.e. door 2 or 3. But in a twist, Monty Hall then, rather than opening any doors, offers you the chance to trade door 1 for doors 2 AND 3, meaning you get to keep whatever's behind both of them. You'd be bonkers to say no; by taking doors 2 and 3 at the same time, you double your chances of winning that elusive car. But this is exactly what happens in a normal situation (assuming you leave behind any goats you win), just with the order reversed.

If that doesn't convince you, then thanks to the small number of objects involved in the problem, we can list all the possible ways the situation can happen.

You choose….	The car is in…	So Monty opens…	If you STICK	If you SWAP
1	1	2 or 3	1 (Car)	2 or 3 (Goat)
1	2	3	1 (Goat)	2 (Car)
1	3	2	1 (Goat)	3 (Car)
2	1	3	2 (Goat)	1 (Car)
2	2	1 or 3	2 (Car)	1 or 3 (Goat)
2	3	1	2 (Goat)	3 (Car)
3	1	2	3 (Goat)	1 (Car)
3	2	1	3 (Goat)	2 (Car)
3	3	1 or 2	3 (Car)	1 or 2 (Goat)

The 'If you SWAP' column is pretty convincing – I've listed every single possible combination of events (known in maths as Proof by Exhaustion) and the majority of the time, if you Swap you win the car. The only times you don't win the car are when you accidentally picked the car as your original choice (which is unlikely, happening only $1/3$ of the time).

I like this problem because it's so counter-intuitive, and emphasises the importance of proof in maths. It's not good enough just to have a statement or idea that seems convincing – you have to prove it's true beyond all doubt, which this table of results does. I also enjoy that fact that in some ways it's the opposite to the problem of how many ways a person can go up a flight of stairs. In that example, the problem was made simpler by lowering the number of steps to one; in this problem it's made simpler by increasing the number of boxes as much as possible.

This problem has two equivalent analogies which I would invite you to try and wrap your head around: The Three Prisoners Problem and Bertrand's Boxes Problem (both included in the appendix).

The Alabama Paradox

On the subject of results which seem counter-intuitive or paradoxical, they don't just happen in hypothetical situations. Allow me to set the scene.

The year is 1880. The US Congress, with the help of a chief clerk C. W. Seaton, was trying to work out how many seats each of the States should get in Congress. The method they used was fairly simple once you get your head around it. Start with a number of seats in Congress (for example 250), and using each State's population, divide it by the total population to work out how many seats each State should get. For example, let's assume that the USA has only three States : Texas (27 people), Alaska (10 people) and Dakota (16 people). That makes the total population of the USA 53 people, so we just divide each State's population by 53. So Texas gets $27 \div 53 \times 250 = 127.3$ seats, Alaska gets $10 \div 53 \times 250 = 47.1$ seats and Dakota gets $16 \div 53 \times 250 = 75.5$ seats

These almost always end up as decimal answers, and so without resorting to allocating parts of senators (best not to think about it), they use a simple system. First, allocate each State as many seats as they can - Texas gets 127, Alaska gets 47 and Dakota gets 75. This has only allocated 249 of our 250 seats, and so the final seat simply goes whoever has the largest decimal left over – in this case, Dakota with 0.5 (against 0.3 and 0.1). After Dakota's extra seat, the final seat allocation is Texas gets 127, Alaska gets 47 and Dakota gets 76.

If there is more than one spare seat, the two highest decimal get them both, and so on. It's a pretty sensible system that appears to be reasonably flawless – the higher your population, the more seats you get, and the more seats there are in total, the more seats you get.

However, this section isn't called the Alabama Paradox for nothing. It turns out that C. W. Seaton was experimenting with different numbers of seats in Congress – he started with 250, and systematically worked his way up to 350, testing out every single number. As you would expect, as he added more and more seats, each state slowly gained more and more seats in Congress. That is, until he reached 299 seats. In 1880, if there had been 299 seats in Congress, Alabama would have been entitled to 8 of them. But increase Congress to 300 seats, and Alabama would only be entitled to 7 of them. For some reason, by increasing the number of seats in Congress, Alabama actually gets less of them.

It's a result that doesn't seem to make any sense, and has been the subject of a lot of study by the mathematical community. C. W. Seaton made Congress aware of this, and 20 years later a new system was adopted, but the name stuck. This paradox is known as an Apportionment Paradox, and can range from allocating teachers to classes to allocating equipment to sports teams. In 1982 (very recent in mathematical terms), it was even proved that Apportionment Paradoxes are unavoidable in this situation, and that it is impossible to devise a system which works perfectly for any number of Seats, states and populations.

It's one of the few times that maths cannot provide a cut-and-dry answer, but I sleep easy in the knowledge that at least maths knows it cannot, and has even gone to the lengths of proving that it cannot. (It's worth noting that proving that something is impossible is much much harder than proving something is possible. It's like trying to prove that there definitely isn't a mouse in your garden – it's much easier to prove that there is a mouse in your garden by simply finding it). And it's not the only case of maths (and mathematicians) knowing that they cannot provide the answers – in 1931, an Austrian mathematician named Kurt Godel proved that any mathematical system you can construct will always contain at least one paradox. Discussions on this topic plunge quickly and deeply into advanced mathematics and philosophy, but it raises the question of the validity of the entire mathematical system; not just the system we use, but any system.

Is a paradox unacceptable for a perfect system of maths? Is a perfect system of maths necessary? These aren't questions I can answer here, and I would argue that perhaps nobody adequately can answer them. Suffice to say that I am very happy with the mathematical system we use, and in my eyes, paradoxes certainly provide more interest than they do inconvenience.

37% of statistics are made up

We all know the phrase "There are lies, damned lies, and statistics". They fill headlines, adverts and political campaigns with claims of varying truthfulness, each with some basis in concrete fact before (some of them) are manipulated beyond recognition.

Imagine there are two political parties, and an election is coming up. Campaigns are running hot, and each party is using any information they can to discredit the other. A talking point is house prices, which both parties feel that the public don't want to see increased. The party currently in power exclaims that house prices have increased just 1% since they took over government, a result of their excellent policies. The opposition derides them, claiming that the house prices have seen a 50% increase due to their extravagant economic policies. Surely one of them has to by lying? There's no way that prices can have increased by 1% and 50%; the figures are too wildly different.

In fact, they are representing the same data, just interpreting it differently (and using sloppy mathematical language). In reality, house price growth had gone from 2% to 3%. It's clear that one could claim that a 1% increase has occurred, but the 50% claim is equally (if not moreso) valid. 50% is a half, and half of 2% is 1%. So the increase of 1% can be expressed as an increase of 50%.

This all seems rather suspicious, but it's the difference between an increase in percentage and an increase in percentage points. In reality, house prices have increased one percentage point, which is an increase of 50%.

Although this is a slightly extreme example, it is one which regularly occurs in the build up to elections – not by the political parties, but by those commenting on it. In a situation that the polls are showing that one political party will receive 40% of the vote, which then increases to 50% of the vote, often commentators will state that the latest polls show a 10% increase for the party's support, but it's not. It's an increase of 10 percentage points, which is actually an increase of 25% (as 10 is a quarter of 40). It may seem a pedantic difference, but when deciding who to choose to lead a country for five years, there's a big difference between 10% and 25%.

Staying on the topic, elections also tend to revolve around debates on spending. The UK currently has a deficit of around £170 billion, which has naturally been the subject of much discussion. The outraged media exclaims that despite this, we still spend £33 million every year on the monarchy! The problem here, and with many other similar claims (depending on your motivations), is that billions and millions are used rather interchangeably as 'large amounts of money'. Especially in televised debates, it can be easy to lose track of the millions and billions, and not really get any sense of the scale of the numbers being used (except that they're very large).

To help our sense of scale, I go back to using time. A million seconds is just over 11 days. A billion seconds is almost 32 years. The two amounts are almost incomparable. To use the figures above (170 billion and 33 million) in seconds, we are comparing 5400 years with 363 days (not quite even 1 year). It's completely insignificant. In fact, I even used rounded figures. It's actually 5387 years, but even those 13 years felt irrelevant.

I'm not saying that £33 million isn't a lot of money, but in the vast majority of cases any number of billions is much, much larger than any number of millions, and it's worth keeping track of that during any discussion about the economy between politicians.

Changing course slightly, we come onto adverts. Let's say that you work in a large office complex, which employs 2000 people. You're the manager of the canteen, and want to find out the office workers' food preferences before you order stock. You could ask every single person, but that's very time consuming, so you just want to ask a selection of people. The question is this – what's the fewest number of people you can ask, and still get reliable data? Clearly if you only ask 5 people, that's not going to be representative; even asking 50 people, you've only got the opinions of

2.5% of the workers, which isn't really that many. I'd plump for a nice round figure of 200 people (10% of the workers). It's probably the smallest number you get away with, as your results will be accurate to within 6% either way – for example, if in truth 70% of people like the canteen, then your results will usually end up between 64% and 76%. It's not concrete, but probably just about reliable enough for most purposes.

Now let's scale up our question – what if you were the CEO of L'Oréal, with a worldwide client base of (I conservatively estimate) 500 million people. What then, if you want to find out whether consumers like your product? What if you want to make claims about the effectiveness of your product, based on real consumers? How many should you ask? Well, going by our previous 10% rule, that's roughly 50 million people – slightly less than the population of England. Now the numbers don't quite scale that simply, and you don't actually need to ask 50 million people, but certainly several thousand responses would be needed.

I watched one particular advert recently, making all sorts of claims about their Revitalift cream and how it made skin look smoother and younger. It was all based on consumer research, and seemed convincing – until you read the little note at the bottom of the screen, saying that they only asked 60 women. Sixty. Not 60 million, not 60 thousand, just 60. It's pathetic. In a customer base that large, it's completely insignificant. It's roughly the equivalent of filling a pot with a million pieces of paper, each with a random number on – then drawing one piece of paper out, seeing it has the number 3 written on it, and then claiming that every single piece of paper in the pot has a number 3 on it.

I understand that asking 50 million people is impractical and expensive – but at least try. There are several adverts for makeup that ask around 1000 people; it's not exactly great, but at least it's a start. It's not 60. In my book, that's intentionally misleading.

However, there are examples of manipulation of statistics much more sinister. One particular incident revolves around gun crime. It's naturally a controversial topic, so let's first get a broader perspective. What does the law say about being the victim of armed crime in general? Well it depends on where you live, and is hugely complicated, but on the whole the law (at least in the UK, and in large parts of the USA) states that you should attempt to escape from the assailant – if that's not possible, then you are allowed to attack them in self defense. It seems sensible, and has some exceptions (for example, police officers on duty have no obligation to retreat) but on the whole it works.

However, in 2005 Florida introduced the Stand Your Ground law, now commonplace across many States in America. It stated that "A person who is not engaged in an unlawful activity and who is attacked … has no duty to retreat and has the right to stand his or her ground and meet force with force, including deadly force…" As a British person, where firearms are generally prohibited anyway, this seems crazy. What if, as a gun-carrying citizen, I wrongly think I'm being attacked, and kill an innocent man? What if someone who sees me defend myself thinks that I am the aggressor, and shoots me to defend themselves? Surely this law will simply increase accidental gun deaths? There are probably arguments to be made on both sides, but rather than write an entire book on gun crime, I'll jump to the maths.

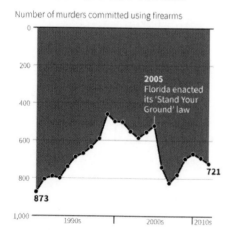

Gun deaths in Florida

Number of murders committed using firearms

2005
Florida enacted its 'Stand Your Ground' law

873

721

1990s   2000s   2010s

Source: Florida Department of Law Enforcement

C. Chan 16/02/2014   REUTERS

Enter Reuters, a reasonably well known news agency which broadcasts and publishes news stories in 10 different languages around the world. In 2014, they published the following graph showing how the number of murders committed using firearms has changed since the Stand Your Ground law was enacted (see left).

When I first saw this, I was surprised – this graph shows that after 2005, the number of murders with firearms sharply decreased, not increased as I thought it would. It took me back slightly. Were the criminals afraid to brandish weapons, knowing that anyone was allowed to shoot them? I wondered at the validity of the data, but the bottom of the graph states that the source is the Florida Department of Law Enforcement, which is probably as reliable as it gets for this sort of thing.

But after looking at the graph for a while, I noticed something. I finally looked at the scale of the vertical axis – it's upside down. 0 is at the top, not at the bottom like it should be. What looks like a decrease in gun deaths is actually a huge increase in gun deaths. It's incredibly misleading, and even as a mathematician it took me a few minutes to realise what was going on. How many thousands of people have seen that graph, and thought that gun deaths had decreased? How many thousands of people could have changed their view on gun laws simply because of a misleading graph? It's a worrying thought. The moral of the story? Always check the scale.

Infinity is bigger than infinity

You've been given a job – write down every positive whole number. As you start (1,2,3,4,5,…) you quickly realise that this is going to take, quite literally, forever. How many positive whole numbers are there? There are, of course, infinitely many.

So before we go any further, what is infinity? Contrary to popular belief, infinity is not a number. Asking whether something is bigger or smaller than infinity is like asking whether something is bigger or smaller than the colour green. There is just no sensible answer to that question. Children arguing will often use infinity to mean 'the most', only to be trumped by the classic "infinity + 1". While I don't blame them, it really doesn't make any sense to say infinity + 1 as it's not a number – again, it makes as much sense as saying green + 1 (i.e. none).

So rather than being a number, infinity is a concept, and specifically in maths it's usually called a *limit*. Nothing can ever reach infinity, but a sequence of numbers can tend towards infinity if it keeps going higher and higher forever. It can get closer to an unattainable goal – sometimes referred to as growing without bound. There's never ever any number you can find which is a maximum cap.

A good example of the idea of infinity is the notion of Hilbert's Hotel, named after mathematician David Hilbert. You are the manager of a fictional hotel, which just so happens to have infinitely many rooms numbered 1,2,3,4,…. and every single room has a guest in it – business is booming. An infinite number of guests are staying in your infinitely many rooms (imagine the queue for breakfast).

The phone rings, and your best mate needs a room for the night. You can't kick out paying guests, so what do you do? It would seem that there's no way to fit him in – until you do something quite clever. Move the guest from Room 1 into Room 2. Move the guest from Room 2 into Room 3. Move Room 3 into Room 4, and so on with every guest in the hotel. Now Room 1 is free, so your friend can stay the night. This isn't an intuitive solution, as in any real hotel you would end up with one guest who was in the 'final room' and couldn't move into the next room – the beauty of infinity is that there is no 'final room', so we can do this without a problem.

The phone rings again – it's not just one person this time, it's infinitely many people and they ALL need a room. Unbelievable. We can't use the same trick as last time, that only creates one room at a time, so to create infinitely many rooms you'd need an infinite amount of time – rather impractical. There is a second trick you can use though - move the guest from Room 1 into Room 2. Move the guest from Room 2 into Room 4. Move Room 3 into Room 6, move Room 4 into Room 8, and so forth. You'll end up with all of the even-numbered rooms with guests in, but all of the odd-numbered rooms are empty. There are infinitely many odd numbers, and so with one swift move you've just created an infinite number of empty rooms, even though you had an infinite number of occupied rooms and couldn't evict any guests.

So you go back to writing down all the positive whole numbers (1,2,3,4,5,…) and even though it would take an infinite amount of time, you could do it. In an infinite amount of time, you could write down all of the positive whole numbers – in maths, this is known as *Countable Infinity* (or listable infinity). It's called Countable Infinity because there are actually two different kinds of infinity – Countable and Uncountable (or listable and unlistable), and Uncountable Infinity is much, much bigger.

A good example of Uncountable Infinity would be to try and write down ALL the numbers between 0 and 1. What would the first number in the list be? 0.1? 0.0001? 0.00000001? There are so many, that you can't even start to list them. Even if you had an infinite amount of time, you couldn't list all of them.

This actually links back to our previous problem about measuring the distance between 3 and 4 (distance 1) – then removing one number at a time (each number having a length of 0) and ending up with nothing left, but still having a measured distance of 1. Because there are an

Uncountable number of numbers between 3 and 4, even with an infinite amount of time, you couldn't remove all of them, which gets us round the problem. You'll always have infinitely many numbers left between 3 and 4, so you can't have nothing with a measured distance of 1.

It seems odd that there is more than one version of infinity, but as shown in the previous paragraph the distinction is vital for advanced maths and our very understanding of numbers.

Some readers may be wondering what sets of numbers are infinitely Countable/Uncountable. One case of particular interest is establishing whether the Rational numbers - that's all the numbers that can be written as fractions, or whose decimals end or repeat - are Countable or Uncountable. (Remember than numbers like 1.5 are rational because they can be written as $^3/_2$). It certainly seems like there are more fractions than there are whole numbers, so Uncountable would seem like a good first guess - but it's wrong. The Rational numbers are actually Countable; there are two methods, which I'll show you.

Remember, to show that a set of numbers is Countable, all we have to do it find a way of listing them (even if writing that list down would take an infinite amount of time).

The first way is simply to make a table of them. Use the row number as the top of the fraction, and use the column number as the bottom of the fraction like this:

$^1/_1$	$^1/_2$	$^1/_3$	$^1/_4$	...
$^2/_1$	$^2/_2$	$^2/_3$	$^2/_4$	...
$^3/_1$	$^3/_2$	$^3/_3$	$^3/_4$	...
$^4/_1$	$^4/_2$	$^4/_3$	$^4/_4$	...
$^5/_1$	$^5/_2$	$^5/_3$	$^5/_4$	...
...	...	...	...	...

This ends up listing every single fraction number – 1.5 is listed ($^3/_2$), and eventually numbers like 2.85 will be listed when the table hits $^{285}/_{100}$, although that's not for a while.

The slight trouble with this method is that it's very inefficient – I suppose when you're talking about an infinite amount of time, it doesn't really matter, but lots of fractions of listed more than once. For example, ½ and $^2/_4$ are both in this part of the table, but they're actually the same number (0.5).

Is there a way of listing every single fraction without repeats? It turns out there is, and it's rather clever. Start with the fraction $^1/_1$. The rule is this: split it into two fractions. One fraction inherits the numerator, one fraction inherits the denominator, and the gaps are filled by adding our current numerator and denominator together. 1+1 = 2, so $^1/_1$ splits like this (the bold numbers are the inherited numbers):

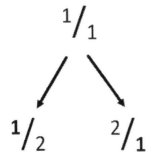

We then continue this process – split each fraction into two fractions, with one inheriting the numerator, one inheriting the denominator, and the gaps being filled by adding the current numerator and denominator.

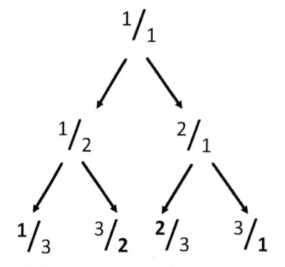

We can then keep this process going forever, and it ends up listing all of the fractions, which was our objective.

What's remarkable is that this never repeats any of the fractions, and never gives two fractions which are equivalent to each other (for example because ½ has been listed, it will never list $^2/_4$ or $^5/_{10}$ or $^8/_{16}$).

So now when someone asks you to list every single fraction in existence, you can make more efficient use of your infinite amount of time – at least it's a start.

Closing thoughts

So here ends our journey through what I think are some interesting and curious elements of mathematics. As promised, at no point did I do any long division or add up big numbers – I'll admit to doing some multiplication, and even writing some fractions, but 2 out of 4 isn't bad.

What have we explored? We have looked at what a number really is, how infinity interacts with our world, how maths is misused and how paradoxes are formed – and taken our first tentative steps into showing how something can be proved. What I really hope to have done is shown you what mathematics truly is, and the vastness of the journeys is can take us on. I've barely begun to scratch the surface of what it's all about, and there's so much maths out there that I don't understand. Frustratingly, I've had to end several sections with a note saying that some of the maths is too lengthy or complex to explain (which it is) – if you're interested to know more, I would encourage you to seek it out. Fully understanding some of the concepts involved is hard work but hugely rewarding. Even if you don't, leave with a sense that THIS is what maths is all about – the very notion of what a number is, proving why a paradox works, and exploring why infinity behaves the way is does.

I've always maintained that a good mathematician is someone who can see patterns, follow rules and produce logical arguments – if you've understood what I've written in this book, then you've certainly done all those things. It doesn't matter if you can't quite remember your 7 times table, by understanding this book you have shown that you are good at maths.

We may have been born too late to explore the earth and too young to explore space, but there really is a vast mathematical world out there waiting to be discovered – it's worth taking a journey there some time. I absolutely love it.

Appendix

As promised, here is the Three Prisoners Problem and Bertrand's Boxes Problem:

Three Prisoners

Three prisoners, A, B and C, are in separate cells and sentenced to death. The governor has selected one of them at random to be pardoned. The warden knows which one is pardoned, but is not allowed to tell. Prisoner A begs the warden to let him know the identity of one of the others who is going to be executed. "If B is to be pardoned, give me C's name. If C is to be pardoned, give me B's name. And if I'm to be pardoned, flip a coin to decide whether to name B or C."

The warden tells A that B is to be executed. Prisoner A is pleased because he believes that his probability of surviving has gone up from $1/3$ to $1/2$, as it is now between him and C. Prisoner A secretly tells C the news, who is also pleased, because he reasons that A still has a chance of $1/3$ to be the pardoned one, but his chance has gone up to $2/3$. What is the correct answer?

Bertrand's Boxes

There are three boxes, each with two coins inside. One box has two gold coins inside (**GG**), one box has two silver coins inside (**SS**), and the other box has one gold coin and one silver coin (**GS**). A box is chosen at random, a random coin is taken out, and it is gold. What is the chance of the other coin in the box also being gold?

Printed in Great Britain
by Amazon.co.uk, Ltd.,
Marston Gate.